discovermore

Earth Science

Landforms

Theia Lake

I0807100

Britannica Educational Publishing

IN ASSOCIATION WITH

THE ROSEN PUBLISHING GROUP

Published in 2026 by Britannica Educational Publishing (a trademark of Encyclopædia Britannica, Inc.) in association with The Rosen Publishing Group, Inc.
2544 Clinton Street, Buffalo, NY 14224

Distributed exclusively by Rosen Publishing.
To see additional Britannica Educational Publishing titles, go to rosenpublishing.com.

Editor: Danielle Haynes
Book Design: Michael Flynn

Photo Credits: Cover Mumemories/Shutterstock.com; (series background) Dai Yim/Shutterstock.com; p. 4 Photoongraphy/Shutterstock.com; p. 5 mapichai/Shutterstock.com; p. 7 (top) Leonid Ikan/Shutterstock.com; p. 7 (bottom) KensCanning/Shutterstock.com; p. 9 (top) jum ruji/Shutterstock.com; p. 9 (bottom) Elena_Suvorova/Shutterstock.com; p. 10 Daniel Prudek/Shutterstock.com; p. 11 RobJ808/Shutterstock.com; p. 13 (top) Zack Frank/Shutterstock.com; p. 13 (bottom) Matthew Dixon/Shutterstock.com; p. 14 Ken Schulze/Shutterstock.com; p. 15 zombiu26/Shutterstock.com; p. 17 (top) Andrei Armiagov/Shutterstock.com; p. 17 (bottom) antony trivet photography/Shutterstock.com; p. 18 Joe Belanger/Shutterstock.com; p. 19 Naeemphotographer2/Shutterstock.com; p. 20 hassan Abbas Ahmed/Shutterstock.com; p. 21 Alex Stemmer/Shutterstock.com; p. 22 Ko Zatu/Shutterstock.com; p. 23 Andreogeno/Shutterstock.com; p. 24 Alexey Fedorenko/Shutterstock.com; p. 25 Heather L. Paisley/Shutterstock.com; p. 26 Andrei Armiagov/Shutterstock.com; p. 27 Jrossphoto/Shutterstock.com; p. 28 Evgeny_V/Shutterstock.com; p. 29 imageBROKER.com/Shutterstock.com.

Library of Congress Cataloging-in-Publication Data

Names: Lake, Theia, author.
Title: Landforms / Theia Lake.
Description: Buffalo, New York : Britannica Educational Publishing, [2026]
| Series: Discover more. Earth science | Includes index.
Identifiers: LCCN 2025000363 (print) | LCCN 2025000364 (ebook) | ISBN 9781641904193 (lib. bdg.) | ISBN 9781641904186 (paperback) | ISBN 9781641904209 (ebook)
Subjects: LCSH: Landforms--Juvenile literature.
Classification: LCC GB406 .L34 2026 (print) | LCC GB406 (ebook) | DDC 551.41--dc23/eng/20250202
LC record available at https://lccn.loc.gov/2025000363
LC ebook record available at https://lccn.loc.gov/2025000364

Manufactured in the United States of America

CPSIA Compliance Information: Batch #CSBRIT26. For further information contact Rosen Publishing at 1-800-237-9932.

Contents

Solid Ground

Only about one-quarter of the Earth's surface is made up of dry land. The rest is covered by water, mostly oceans. Even though the amount of land is small compared to the water, it's varied. There are huge mountains, dark caves, hills of sand in the desert, and islands of all sizes. These and other features, shapes, and types of land are called landforms.

From outer space, it's easy to see how oceans cover most of the Earth. Humans rely on both the land and the water.

Underwater volcanoes in the sea leave lava on the seabed. It gradually builds up until the volcano rises above the surface, forming an island.

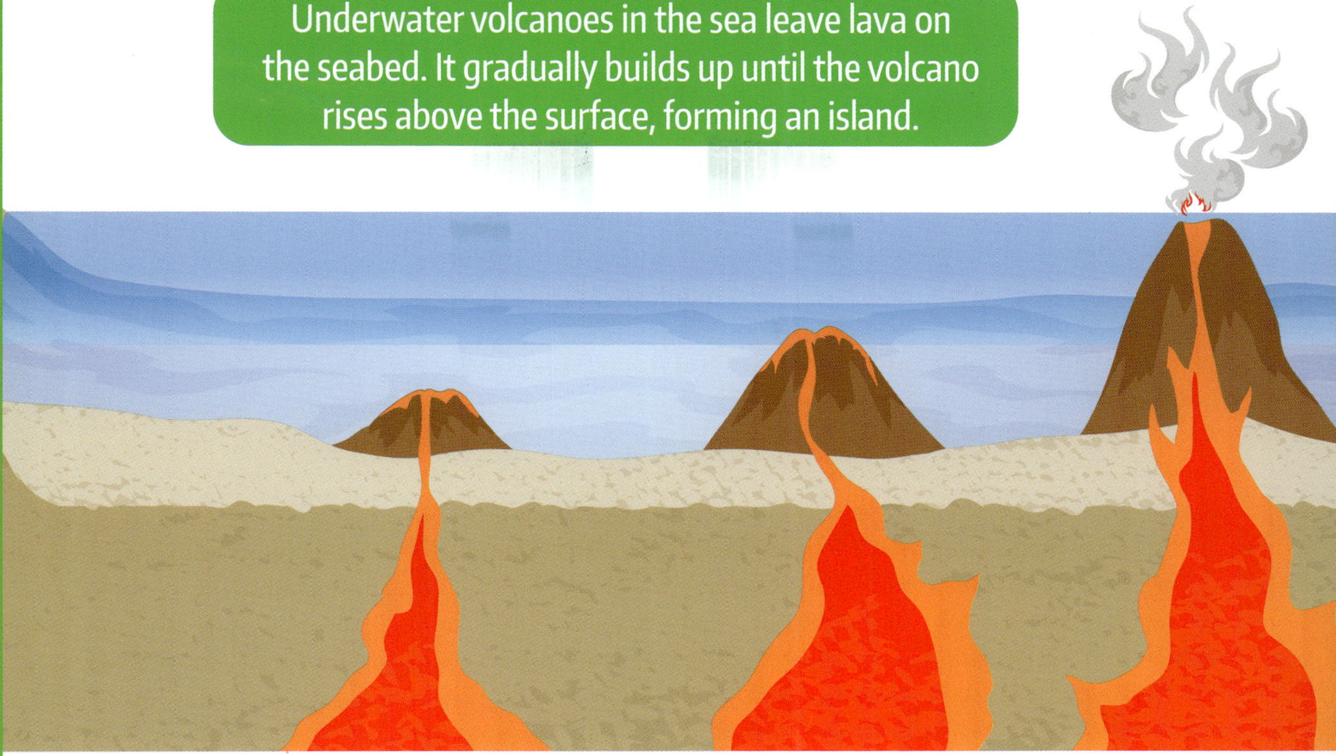

Landforms vary around the world. Australia has large areas of sandy desert, and Nepal has many mountains. Most landforms form slowly, for example, when rock is gradually worn away by **erosion**. Some landforms form more quickly, for example, when hot lava from underground bursts out onto the surface and hardens. This is how many islands are formed.

WORD WISE

EROSION IS WHEN SOIL OR ROCK IS CARRIED AWAY BY WATER, ICE, OR WIND.

Vast Plains

Plains are a type of landform that are generally flat or have gently rolling hills without trees. These common landforms cover around one-third of all land on Earth. They are found on all continents except Antarctica. Many plains are covered by grassland. These are areas where mainly grass grows and animals such as zebra, deer, and bison live.

Plains form in different ways. Some are formed by coasts when rivers slow down as they meet the sea. The tiny pieces of rock the rivers erode stop moving, drop, and pile up in flat plains surrounding the river. They can also form in areas where the erosion of sediments from hills or mountains pile up on lower-lying land.

The biggest plain landform in the world is the West Siberian Plain in central Russia. It has poor drainage so there are many swamps in this plain.

compare and contrast

Many plains animals have hooves. Compare and contrast hooves and other shapes of feet. How are hooves useful for life on the plains?

Most of the vegetation in the Great Plains of the United States is grassland.

Plateaus, Buttes, and Mesas

Plateaus are also flat, like plains, but they're usually higher in elevation. They range in size from a few yards to hundreds of miles wide. Some plateaus formed when land was pushed upward. The Tibetan Plateau rose when two enormous slabs of rock underground pushed into each other, forming a flat, raised area.

Other plateaus formed when layer upon layer of lava built up. Many plateaus formed when existing mountains and plains were eroded at their tops or edges. Flat-topped hills with steep sides are called mesas. Smaller plateaus are called buttes. There are many mesas and buttes in the southwestern United States, such as in Monument Valley.

Large plateaus often feature an escarpment, a steep slope or cliff, where it meets lower-elevation areas.

The largest plateau in the world is the Tibetan Plateau in Asia. It's surrounded by the tallest mountains in the world.

compare and contrast

How are plateaus and plains similar? How is the height of the surrounding land different?

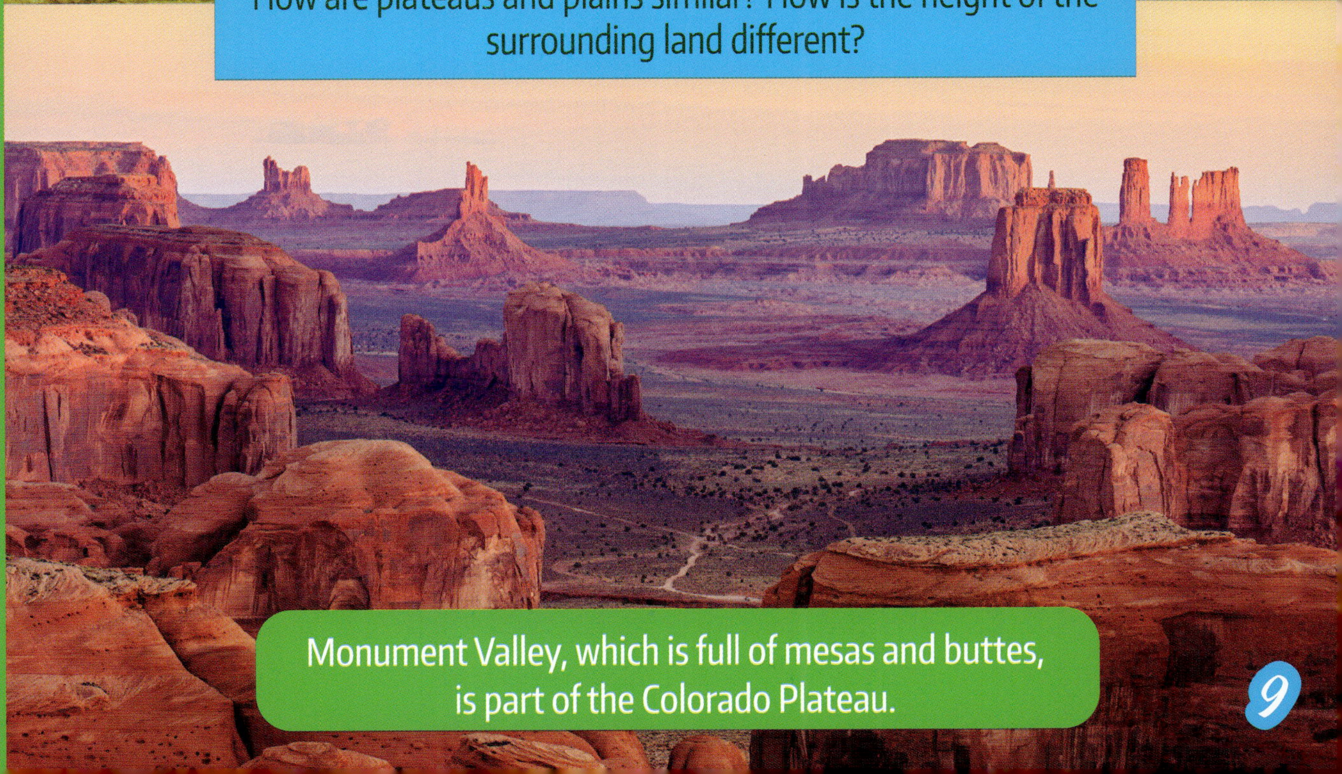

Monument Valley, which is full of mesas and buttes, is part of the Colorado Plateau.

Towering Mountains

Mountains are the tallest and perhaps most awe-inspiring landforms on Earth. Many have a pointed or jagged peak. Some are in rows called mountain ranges. The Himalayas mountain range includes Mount Everest. Its peak is more than 5 miles (8 km) above sea level, making it the highest mountain in the world. It formed when underground slabs of rock collided, making surface rock crumple into folds.

Most of the Himalayas mountains—including Mount Everest in center—are in Bhutan, China, India, and Nepal.

Mauna Kea is a shield volcano, though its last eruption was about 4,500 years ago.

Some mountains form above cracks in Earth's surface. The rock on one side of the crack can rise above the other. Other mountains form when volcanoes erupt many times and layers of hard lava build up. The tallest mountain on Earth from base to peak is a volcanic mountain called Mauna Kea in Hawaii. It is more than 6 miles (9.6 km) from the ocean floor to its peak, but reaches only 2.6 miles (4.2 km) above sea level.

How Hills Form

Like mountains, hills are raised landforms formed in many different ways. Hills are not as high as mountains, though there is no official difference between the two. Some hills formed when rocks collapsed or when blocks of land rose. Others grew when magma, or other liquid masses, pushed up from under the ground, causing the land to bulge.

Hills also form by erosion. Sometimes ancient mountain rock is broken down over thousands of years by the action of ice forming on rock and then cracking it open. Rain and wind carry pieces of rock away and the mountain shrinks. The small bits of rock and soil created by erosion can also form hills when they pile up in new places.

Some hills, such as the Black Hills of South Dakota, form when the land lifts up over time.

compare and contrast

How are hills and mountains similar? How are they different?

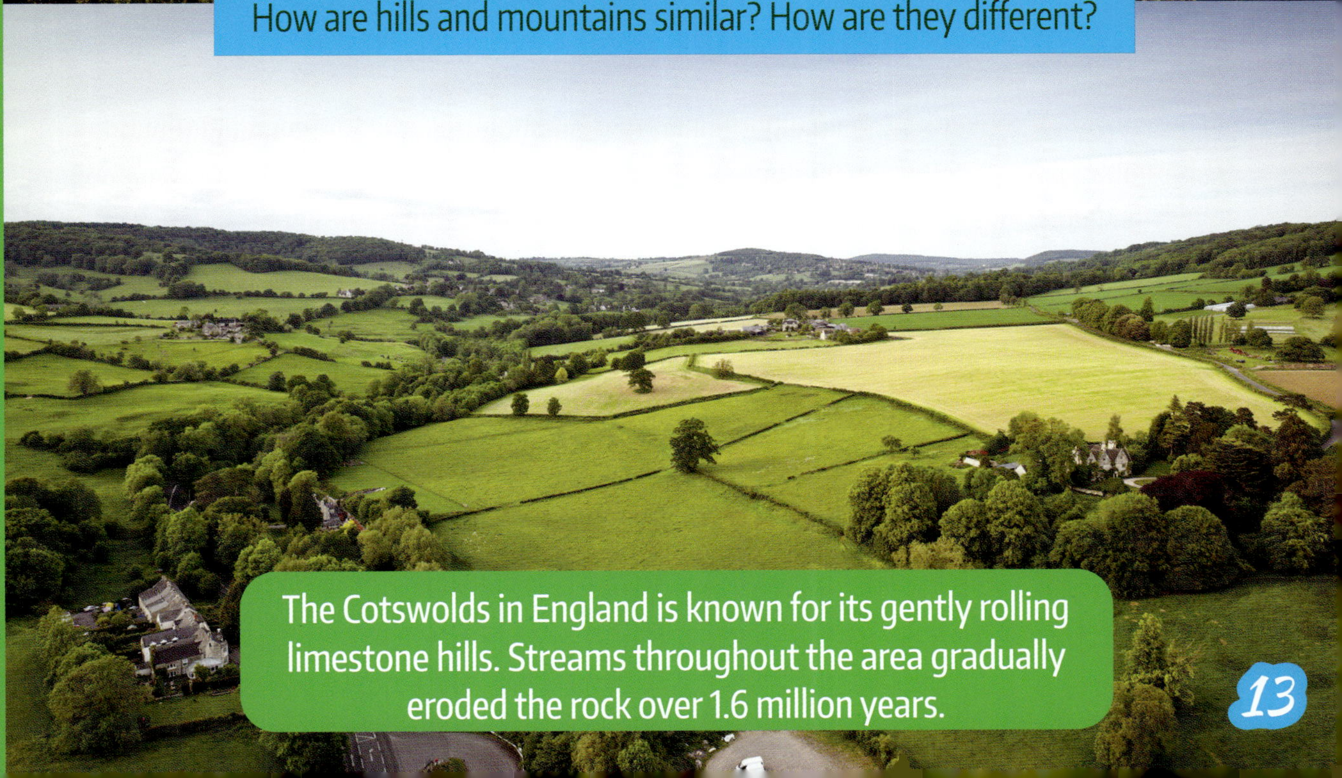

The Cotswolds in England is known for its gently rolling limestone hills. Streams throughout the area gradually eroded the rock over 1.6 million years.

Slow-Moving Glaciers

Glaciers aren't actually landforms, but they play an important role in creating landforms. Glaciers are large areas of thick ice that remain frozen from one year to the next. Glaciers also slowly flow over the land. They exist all over the world, from Antarctica to mountainous points along the equator.

Glaciers form when layers of snow and ice collect in gaps between high mountains and freeze into a solid lump of ice.

Some glaciers melt and break apart when they reach a body of water, such as Margerie Glacier, pictured, where it meets the waters of Glacier Bay in Alaska.

Glaciers slowly break away rocks and drag them down mountains as they move toward their terminus, or end. Those small rocks then gather as sediment at this terminus.

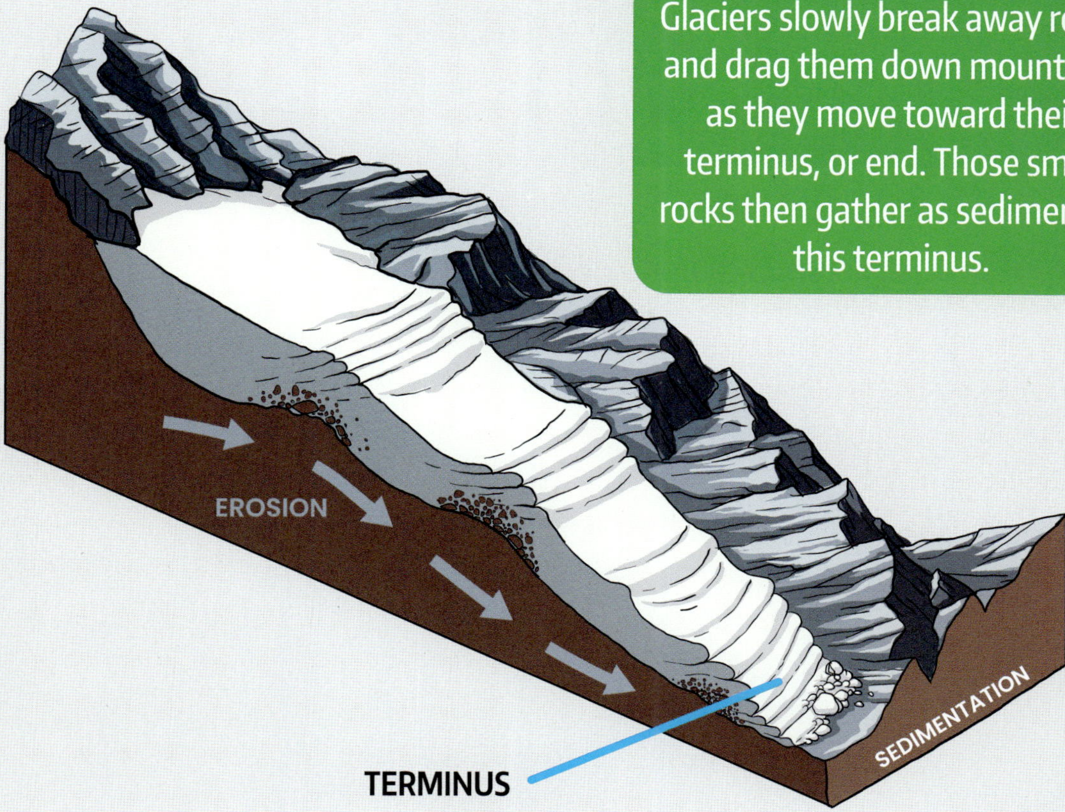

Glaciers wear down mountains. Their edges freeze onto and then pull off chunks of rock as they pass. They push along rock and soil in front of them. When the front edge of the glacier reaches warmer water or its terminus, it melts or calves, which means to break apart.

Consider This

The speed at which glaciers move depends on how much water they slide along on. How does a warmer climate affect glaciers?

Low-Lying Valleys

Valleys are low areas of land between hills or mountains. Most valleys form where water wears away rock or soil.

Rainwater runs off slopes into grooves, forming streams. Streams wash away soil, making a narrow, shallow valley. Over time, the streams turn into rivers, with more water to erode sediment from the valley sides and base. The valley then gets deeper and wider.

In cold places, glaciers can form in river valleys. They erode these valleys into wider glacial valleys. Other valleys, such as the Great Rift Valley in Africa, formed when tectonic plates that make up the Earth's surface began to pull away from each other.

Some valleys have dry land, while others feature a river or lake. Here, the Lodalen Valley in Norway is home to Lovatnet Lake.

compareandcontrast

How are the shapes of valleys made by rivers and those made by glaciers similar? How are they different?

The Great Rift Valley in Africa is enormous, nearly 300 miles (480 km) at its widest point. It formed when tectonic plates began pulling away from each other.

Canyons and Gorges

One of the most dramatic kinds of valleys in the world are canyons. They are especially deep, narrow valleys often with steep, high sides. One of the biggest of them all is the Grand Canyon. This remarkable landform has taken six million years to form. It is the result of the Colorado River eroding the Colorado Plateau.

The Colorado River carved out the Grand Canyon. Some areas, such as Horseshoe Bend, offer stunning views.

Yarlung Tsangpo River, which formed a gorge in the Himalayas mountains, is 1,800 miles (3,000 km) long. It flows through the Tibetan Plateau and has an average elevation of 13,000 feet (4,000 m).

Canyons with very steep sides are called gorges. They form where rivers carve through weak areas of rock. One of the largest gorges in the world is Yarlung Tsangpo Gorge in Tibet. It's three times deeper than the Grand Canyon!

Ever-Changing Sand Dunes

Wind and flowing water are key in the formation of sand dunes. Sand is made up of tiny pieces of rock eroded from large rocks. Sand is light, so wind can blow it around. In places with lots of sand, such as big beaches or deserts, wind piles sand up into hills called sand dunes. The biggest sand dunes are hundreds of yards high.

Wind is constantly changing the formation of sand dunes. Next year, this same area of the Sahara Desert in Egypt may look totally different.

Plants keep the wind from blowing sand away too fast, allowing dunes to form and grow bigger.

Sand dunes grow and change over time. The side facing the wind has a gradual slope. It builds up as wind moves the sand grains on top of other grains. Once the dune becomes too steep, one side may fall, leaving a steeper slope. The tallest sand dune in the world is the Duna Federico Kirbus in Argentina.

Consider This

Some coastal dunes form only where there are plants growing. How do you think these help the sand to pile up?

Let's Explore Caves

Caves are landforms that form underground. They are holes that usually form in limestone. First, rainwater gets underground by soaking into rocks and through cracks. Then it wears away limestone, making gaps that widen over time to form caves. Caves range in size from tiny passages to huge systems of connected "rooms" and tunnels. The world's longest cave system is in Kentucky. It is more than 350 miles (560 km) long.

Kentucky's enormous Mammoth Cave is so impressive, the U.S. government made it into a national park.

Stalactites and stalagmites often grow in the same areas, such as in the Frasassi Caves in Italy.

Inside limestone caves, there are many special features formed by minerals in water. Minerals in water that drips into the cave can build up into shapes called **stalactites**, which hang from the roof. Other forms, called **stalagmites**, grow up where the water splashes onto the floor! When visiting a cave, it's very important not to touch stalactites or stalagmites because the oil from your hands can damage them.

WORD WISE

STALACTITES ARE ROCKY COLUMNS THAT GROW DOWN FROM A CAVE ROOF. STALAGMITES ARE ROCKY PILLARS THAT GROW UP FROM A CAVE FLOOR.

Coastlines and Spits

Every major body of water—oceans and lakes—have a coastline where the water meets land. They are shaped by waves. Waves hit rocks with great force, break off pieces, and wash them away. They force air into cracks and split open rocks. Waves also break rocks when they toss sand and stones against them. Waves erode coastlines to make features such as cliffs and coves, which look like giant bites in the coastline.

One of the most famous coastlines is the White Cliffs of Dover in England. These cliffs are mostly made of soft, white chalk, which erodes easily when hit by waves.

The spit in Homer, Alaska, is so large people have built homes and businesses on it.

Coastlines are also formed when waves drop the sediment they are carrying at the edge of land. Beaches form in bays where sand and broken shells collect. Spits are long ridges of sand or rock pieces that stick out from the coastline into the sea. The Homer Spit in Alaska is 4.5 miles (7.3 km) long!

Consider This

The shape of coastal landforms depends on how hard the rock is. Why do you think a bay forms in an area of soft rock?

Islands of All Sizes

Islands are pieces of land that are smaller than a continent and completely surrounded by water.

They can be very small or as large as Greenland and Australia. Islands form in different ways. Some islands were once attached to a larger area of land but were separated by land erosion. Other islands are volcanoes that erupted underwater. Gradually, the lava builds up until the top of the volcano reaches above the water to form an island.

Many islands grouped together, such as Henningsvaer Lofoten in Norway, are called an archipelago.

You can walk or drive to Mont-Saint-Michel when the tide is out, but make sure to keep an eye on the tide!

Some islands are only islands for part of the day! When the **tide** is out, you can walk across land to these islands. When the tide comes in again, water covers the land between the two, and the island is completely surrounded by water. Mont-Saint-Michel in France and St. Michael's Mount in England are famous temporary islands.

WORD WISE

A TIDE IS THE RISING AND FALLING OF THE OCEAN.

Protecting Landforms

It's not just sand dunes that look different over time. All landforms change, though at different speeds. Some change very slowly, over thousands of years, as they erode. Others change more quickly. People change landforms when they use machines to dig holes in Earth's surface to get coal or other useful substances. The rocks and soil they dig out are often piled up to create new hills, too.

Miners created this large quarry, or open pit, in Russia to get to valuable resources.

Governments create nature preserves, such as Katmai National Park in Alaska, to protect the natural beauty and habitats of wildlife.

Sometimes people create new land. The southern tip of Manhattan, New York City, was extended by landfill to increase the size of the island. In the Netherlands, water was pumped out of marshy land so people could live and farm on it. People also protect landforms to stop them from changing. National parks are places where people are not allowed to alter the landforms.

Consider This

Why do you think people can change landforms more quickly today than they could in the past?

Glossary

Antarctica The very cold region around Earth's South Pole.
collapse To fall apart or down.
collide To push into each other.
continent A large, solid area of land.
cove Small, sheltered area that is set back from the shoreline.
elevation The height above sea level of something.
eruption When a volcano explodes and hot, melted rock, called lava, spurts out of it.
landfill Waste that is buried between layers of earth.
lava Melted rock from inside Earth that comes to the surface when a volcano erupts.
limestone A type of soft rock that erodes easily.
magma Melted rock that is found below Earth's surface.
peak The top of a mountain or hill.
sediment Tiny pieces of rock, mud, dust, or sand.
swamp A wetland with some vegetation.
varied Diverse of different.
volcano Vent in Earth's surface from which melted rock, called lava, spurts out.

For More Information

Books

Hicks, Dwayne. *That's a Valley!* New York, NY: Gareth Stevens, 2022.

Lundgren, Julie K. *Landforms and How They Are Made.* New York, NY: Crabtree Publishing, 2022.

Sohn, Emily, and Adam Harter. *Landforms.* Chicago, IL: Norwood House Press, 2020.

Websites

Britannica Kids: Landform
www.kids.britannica.com/kids/article/landform/599368
This Britannica Kids website shares more information about all kinds of landforms, such as coastlines, glaciers, islands, and valleys.

National Geographic: Landforms and Landscapes
www.education.nationalgeographic.org/resource/resource-library-landforms-and-landscapes/?page=1
Learn more about the land all around us at the National Geographic website.

Index